FLORA OF TROPICAL EAST AFRICA

RESTIONACEAE

HENK BEENTJE[1]

Evergreen rush-like plants, usually dioecious; stems (culms) erect, simple or branched, photosynthetic. Leaves generally reduced to sheaths which are split to their base, usually with a small awn or mucro. Inflorescence sexually differentiated or the sexes similar, terminal, spicate or paniculate, much or little branched. Flowers nearly always aggregated into spikelets, these surrounded at base by a spathaceous sheath, each flower in the axil of a bract, sessile or pedicellate. Flowers small, wind-pollinated, regular; perianth of 2 whorls of 3 segments each, the segments similar or differentiated. Male flowers with three 1-celled dorsifix anthers, introrse with a longitudinal slit, and usually with vestigial female parts. Female flowers with the ovary superior, with 1–3 uniovular locules and 1–3 styles; staminodes 3 or absent. Fruit small, a 1–3-locular capsule or a 1-locular nut. Seeds 1–3, with copious endosperm; embryo small.

The Restionaceae has some 55 genera and 490 species, mostly in South Africa and S & SW Australia, with only a few in Africa, Madagascar, Indo-China and Chile.

RESTIO

Rottb., Descr. Pl. Rar., Progr.: 9 (1772); Linder in Bothalia 15: 64 (1984); Linder et al. in Kubitzki, Fam. Gen. Pl. Vasc. 4: 435 (1998)

Tufted, rhizomatous or mat-forming. Stems of ♀ as in ♂, branching, rarely simple, cylindric or compressed to tetragonal; sheaths persistent. Male flowers aggregated into spikelets; bracts cartilaginous. Female inflorescences similar to the male, but spikelets usually fewer and larger. Perianth cartilaginous or bony, the outer lateral tepals glabrous or villous, keeled. Styles 3; ovary 1–3-locular. Fruit usually dehiscent.

93 species, mostly in the Cape but a few in tropical Africa and Madagascar.

Restio mahonii (*N.E.Br.*) *Pillans* in Trans. Roy. Soc. S. Afr. 30: 255 (1901), as *mahoni*; Linder in Bothalia 15: 452 (1984) & in K.B. 41: 99 (1988). Type: Malawi, Zomba, *Mahon* s.n. (K!, holo., B, BOL, iso.)

Plants tufted, often forming cushions or mats to 1 m deep. Culms branching, cylindric, 30–60(–200) cm high, 1–2(–3) mm in diameter at base, branched from the middle, glabrous, rugulose, the terminal branchlets sometimes compressed; sheaths tubular, 8–20 mm long, with broad hyaline upper margins, ending in an awn to 10 mm long (but mostly broken and appearing less than 1 mm long in dried material). Inflorescences of ♂ plants composed of 1–8 spikelets arranged in small racemes 2.5–4 cm long, those of ♀ plants of 1–4 spikelets; lowermost spikelets distant. Spikelets

[1] This treatment is based on the work of H. Peter Linder in Bothalia and Kew Bulletin. I would like to thank Dr Linder for his comments on the manuscript.

7
5
6
4
2
3
JW
1

oblong, 5–10 mm long, 3–4 mm in diameter, hardly compressed, 1–8-flowered. Spathe similar to the sheath, 5–12 mm long, 2–2.5 mm wide, shortly mucronate. Bracts loosely embracing the flowers, lanceolate, 4–9 mm long, acute to mucronate, membranous on the upper margins. Flowers shortly pedicellate, the pedicels < 1 mm long, villous. Perianth segments lanceolate to triangular, 3–5 mm long, cartilaginous, glabrous, the dorsal tepal flat and acute, the laterals conduplicate and very acute, puberulous outside; inner segments slightly ovate-lanceolate, 2.5–4 mm long, 1–1.5 mm wide, acute. Male flowers: stamens with flattened filament, the anther oblong, 1.5–3 mm long, acute, exserted at anthesis; pistillode very short, topped by 3 style rudiments. Female flowers: ovary globose, 2-chambered, with prominent lateral sutures; styles 3, widely spaced, plumose; staminodes 3, ± 1 mm. Fruit a capsule, almost lenticular, 2–2.5 mm long, ± 2 mm in diameter, with two 1-seeded locules, coriaceous; seed triangular in section. Fig. 1.

subsp. **mahonii**; Linder in K.B. 41: 100 (1988)

Inflorescence of racemosely arranged spikelets; flowers nearly always unisexual.

TANZANIA. Morogoro District: Uluguru [Ourougourou] Mts, Mt Makumbaku [Mkombako], 1913, *Sacleux* 2611!

DISTR. **T** 6; Congo (Kinshasa), Malawi, Zimbabwe, Madagascar

HAB. Not mentioned for our single specimen (elsewhere in peaty or marshy habitats, 1500–3000 m); Mt Makumbaku reaches 2420 m

USES. None recorded for our area

CONSERVATION NOTES. Fairly widespread, but in Congo only known from one volcanic peak, in Tanzania not collected for 90 years, status in Madagascar and Zimbabwe unknown, common on Mt Mlanje (Linder, pers. comm.); probably Least Concern (LC)

SYN. *Hypolaena mahonii* N.E.Br. in F.T.A. 8: 265 (1901), as *mahoni*; Pillans in Trans. Roy. Soc. S. Afr. 16: 398 (1928)

Restio madagascariensis Cherm. in Bull. Soc. Bot. Fr. 69: 318 (1922). Type: Madagascar, Mt Ibity, *Perrier* 2785 (P, lecto., chosen by Linder)

NOTE. The other subspecies, subsp. *humbertii* (Cherm.) Linder, is restricted to Madagascar and differs in the capitate inflorescence and usually bisexual flowers.

FIG. 1. *RESTIO MAHONII* — **1**, habit, female plant, × $^1/_2$; **2**, male flower spikelet, × 3; **3**, male flower, × 6; **4**, male flower, anthers and pistillode rudiments × 6; **5**, female flower spikelet, × 3; **6**, female flower, × 6; **7**, fruit, × 6. All from *Robinson* 5291 & 5292. Drawn by Juliet Williamson.

INDEX TO RESTIONACEAE

There are no new names validated in this part

First published in 2005 by
Royal Botanic Gardens, Kew
Richmond, Surrey, TW9 3AB, UK
www.kew.org

ISBN 1 84246 064 1

Design by Media Resources, typesetting and page layout by Margaret Newman,
Information Services Department,
Royal Botanic Gardens, Kew.

Printed by Cromwell Press Ltd.

For information or to purchase all Kew titles please visit
www.kewbooks.com or email publishing@kew.org